Cool and Crazy
Engine Coloring Book

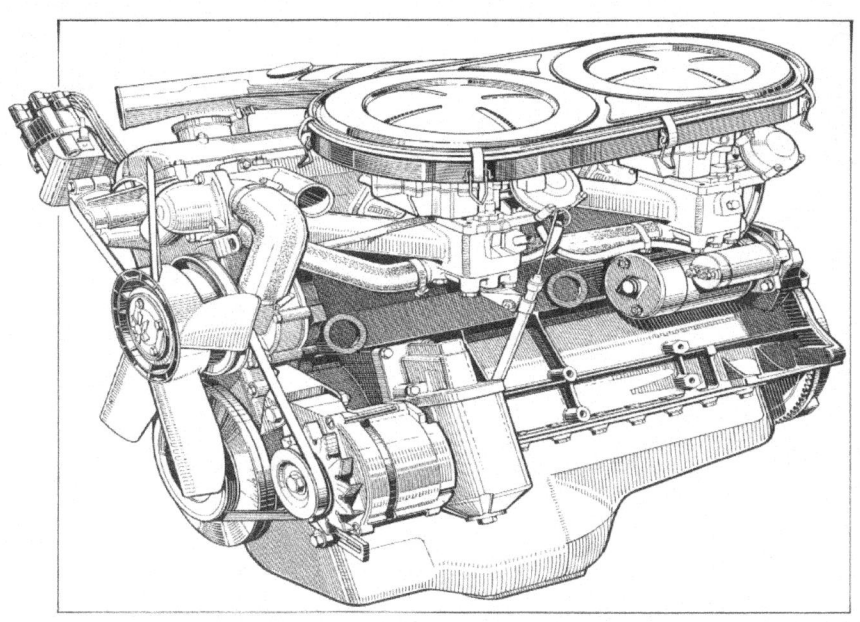

Detailed Exploded and Cut away
diagrams of Combustion Engines.

To COLOR

By

NailHeadHotRods

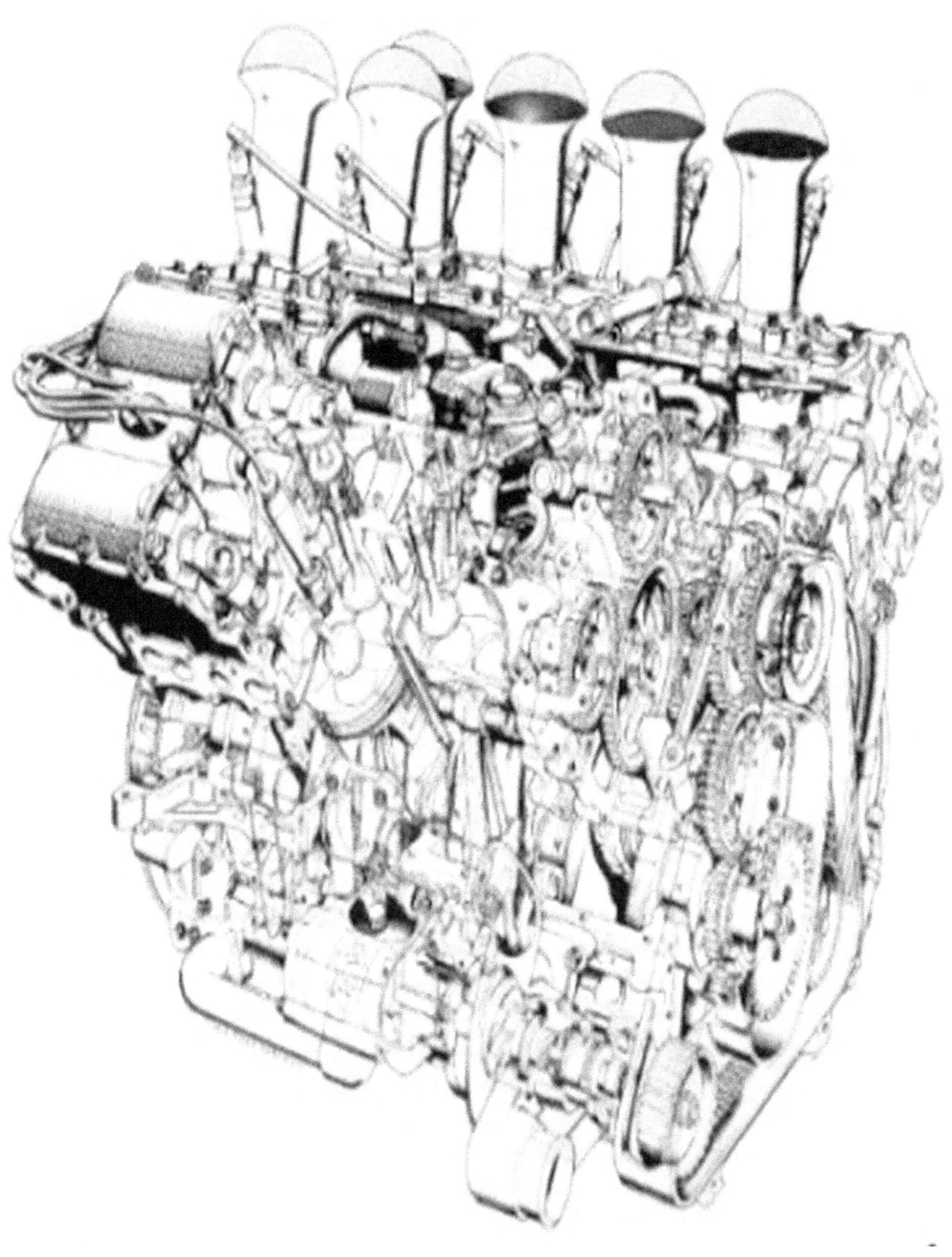

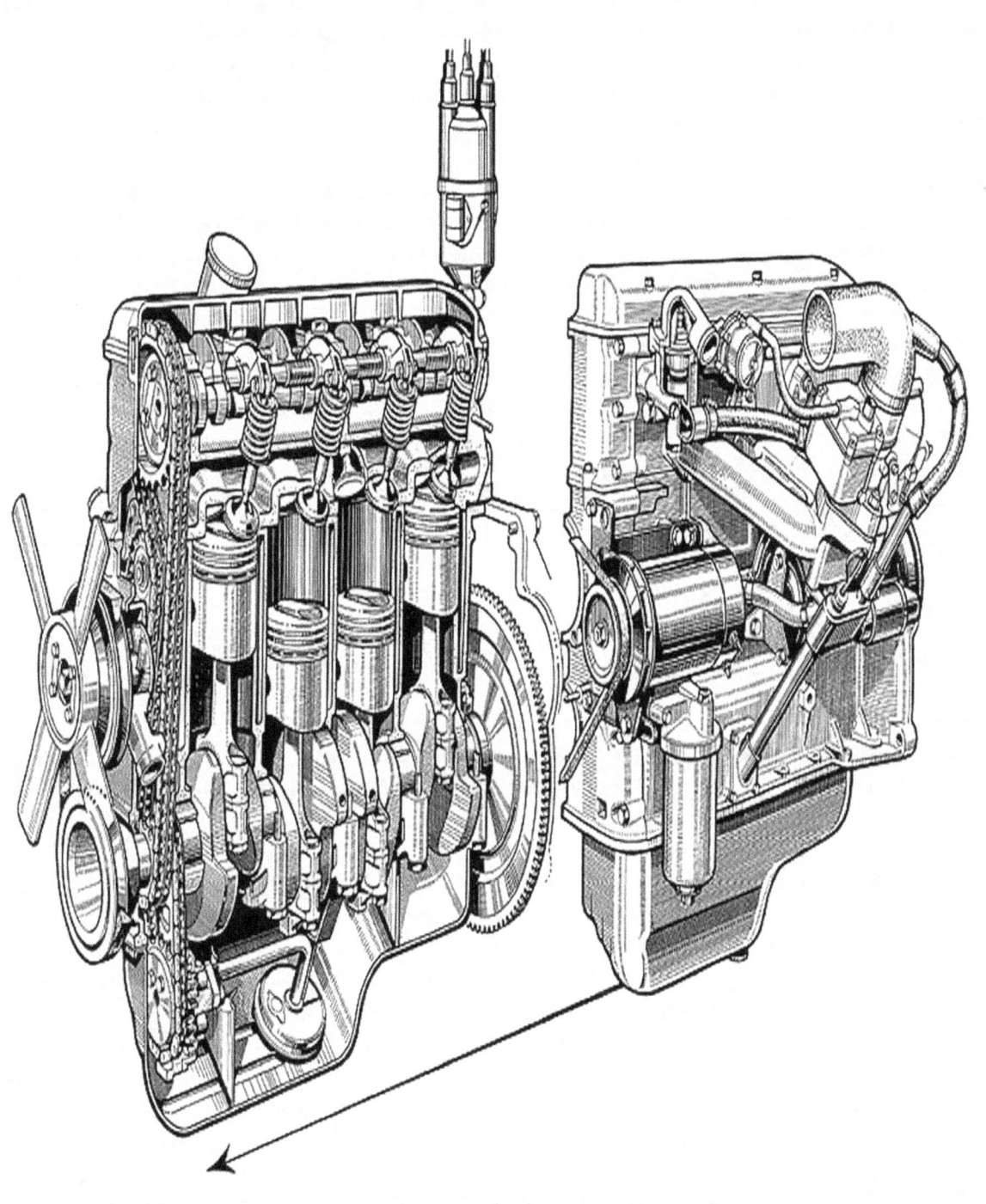

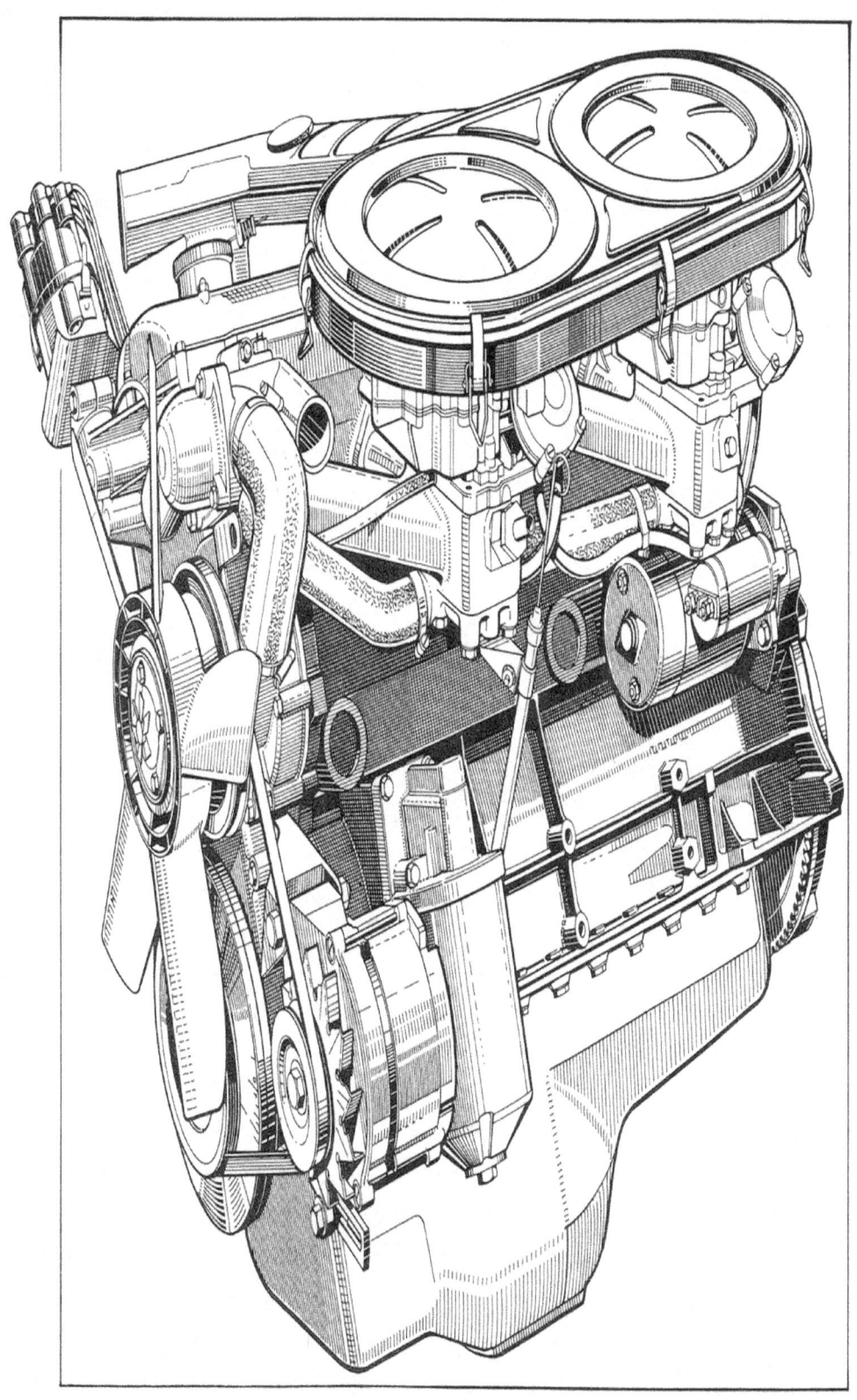

The Internal Combustion Engine

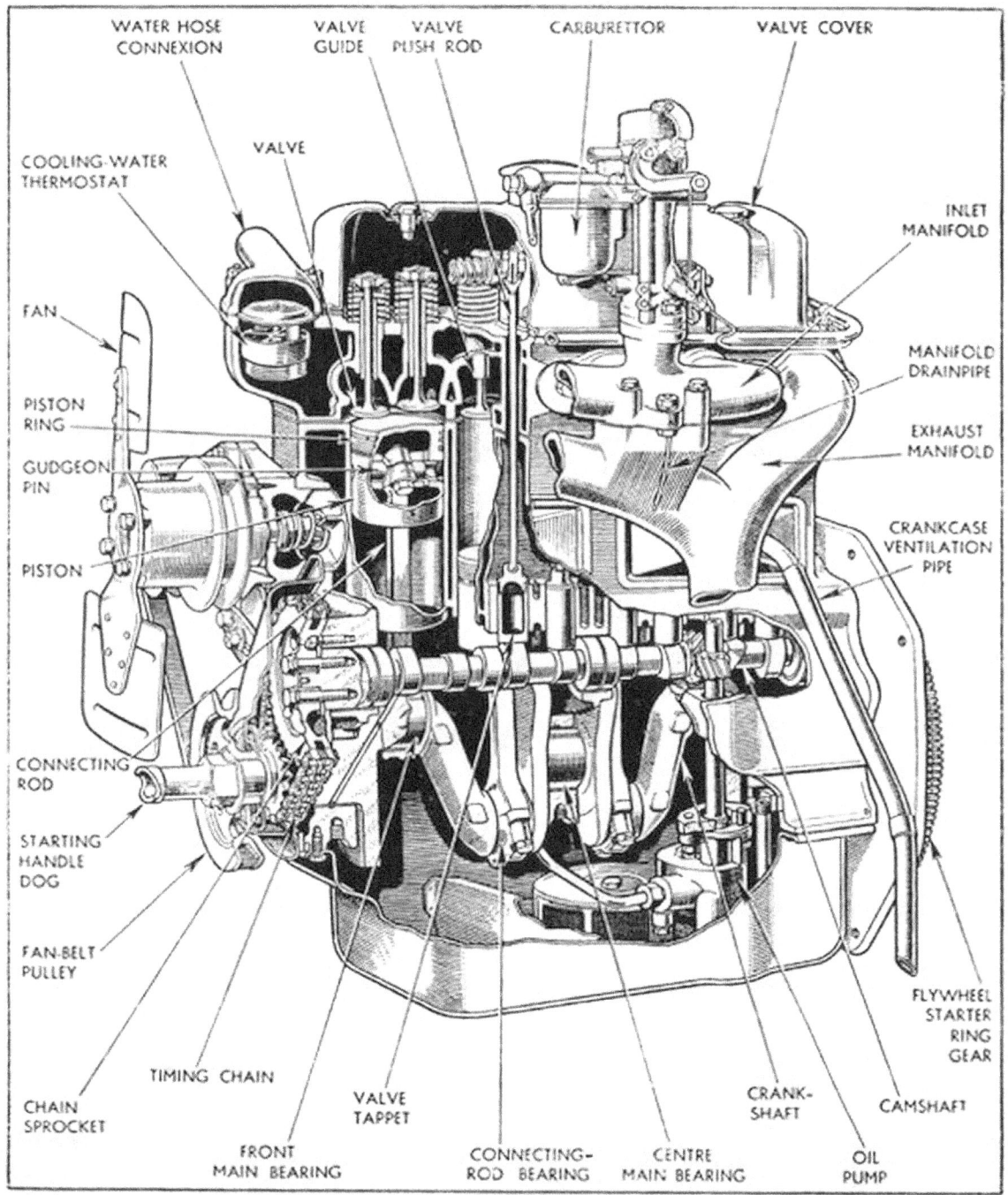

WATER HOSE CONNEXION · VALVE GUIDE · VALVE PUSH ROD · CARBURETTOR · VALVE COVER · COOLING-WATER THERMOSTAT · VALVE · FAN · PISTON RING · GUDGEON PIN · PISTON · CONNECTING ROD · STARTING HANDLE DOG · FAN-BELT PULLEY · CHAIN SPROCKET · TIMING CHAIN · FRONT MAIN BEARING · VALVE TAPPET · CONNECTING-ROD BEARING · CENTRE MAIN BEARING · CRANK-SHAFT · OIL PUMP · CAMSHAFT · FLYWHEEL STARTER RING GEAR · CRANKCASE VENTILATION PIPE · EXHAUST MANIFOLD · MANIFOLD DRAINPIPE · INLET MANIFOLD

This partly cut-away view of a modern four-cylinder power unit illustrates most of the components. The main features can be quickly identified; the crankshaft which drives the camshaft through a double roller-chain, the pistons, valves and valve gear, the inlet and exhaust systems, and the oil pump which forces oil under pressure through an intricate system of oil-ways in the engine. The only items not shown are the col-ignition distributor and the sparking plugs.

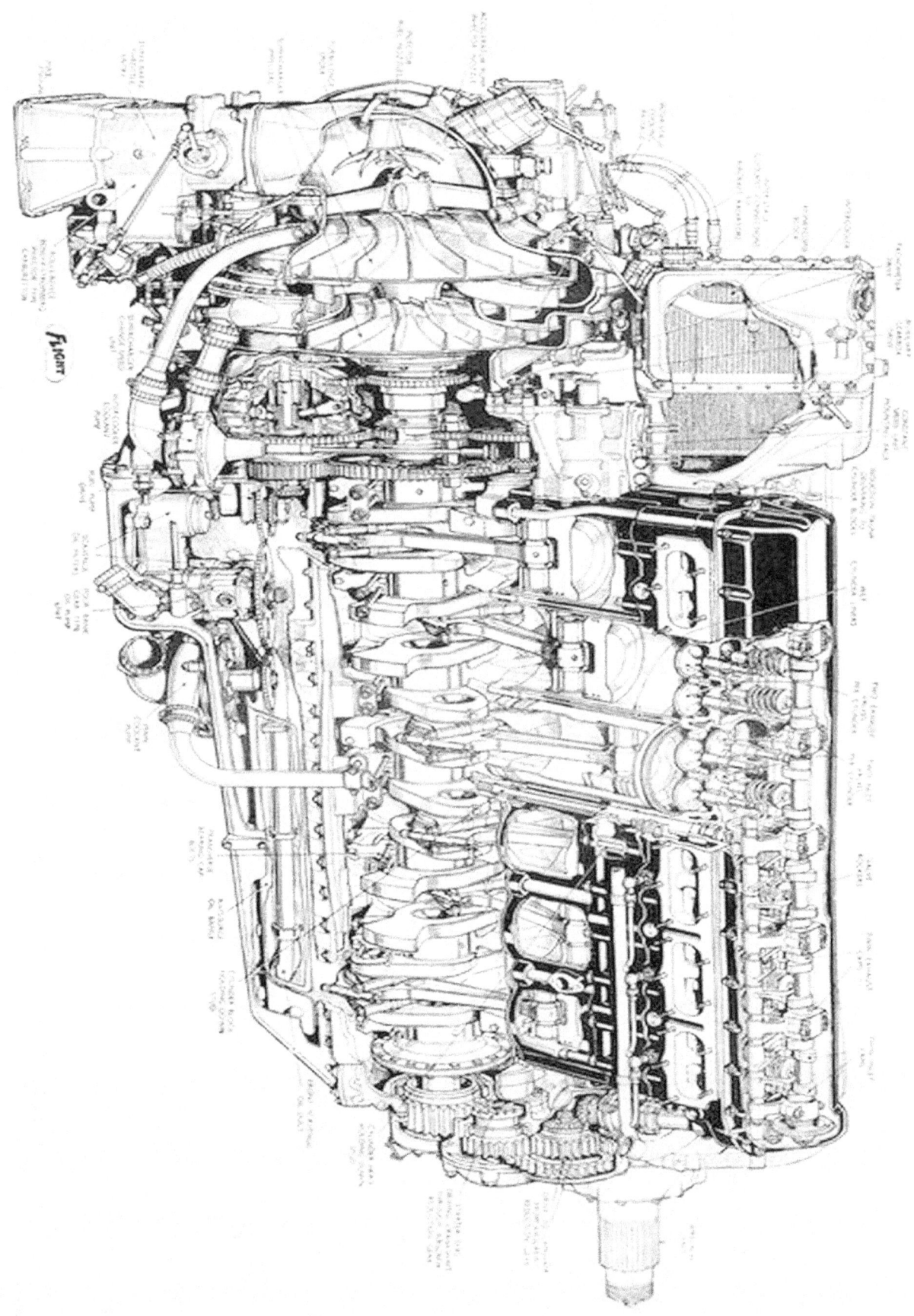

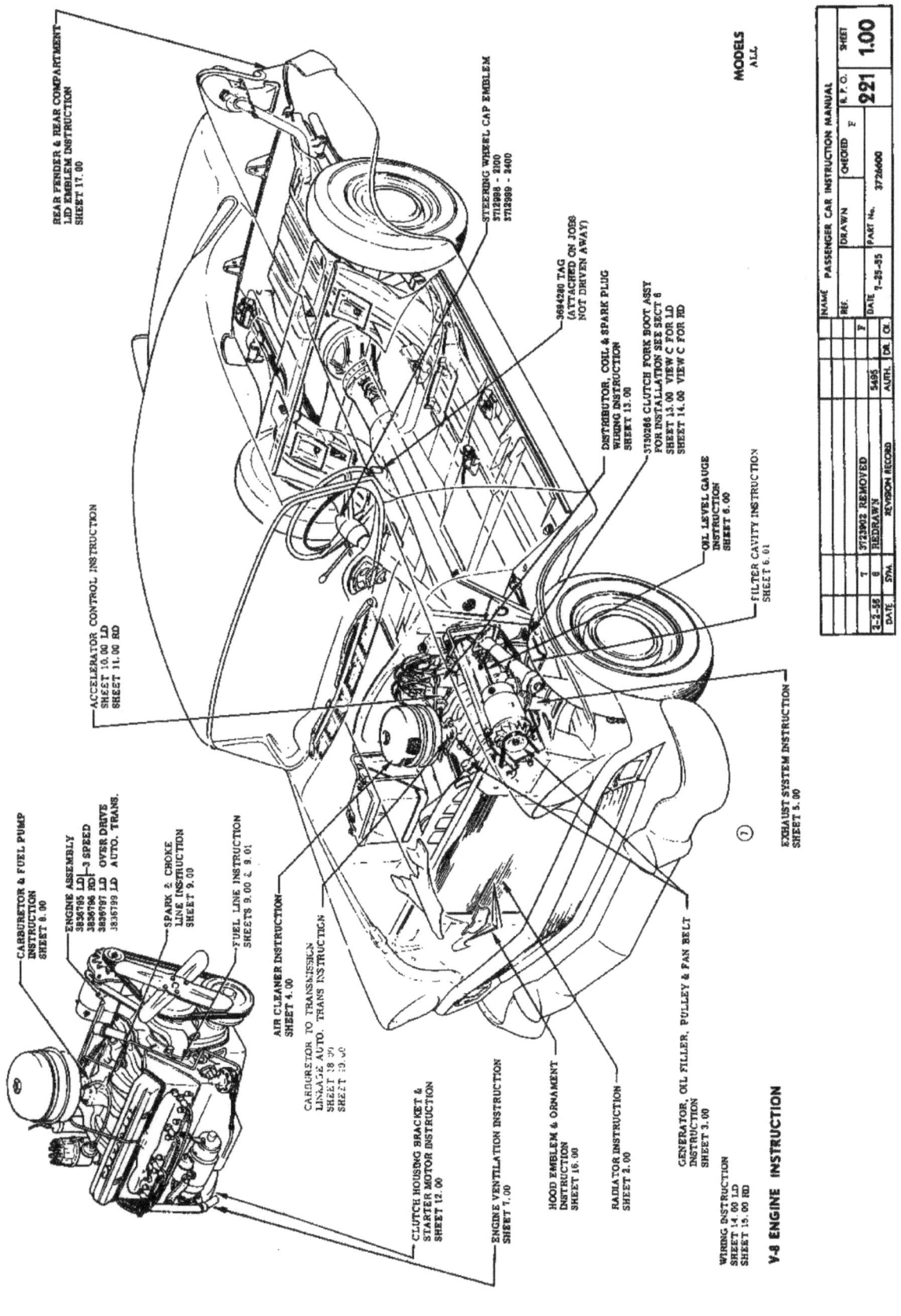

REAR FENDER & REAR COMPARTMENT
LID EMBLEM INSTRUCTION
SHEET 17.00

STEERING WHEEL CAP EMBLEM
3712985 - 2100
3712989 - 2400

3694280 TAG
(ATTACHED ON JOBS
NOT DRIVEN AWAY)

DISTRIBUTOR, COIL & SPARK PLUG
WIRING INSTRUCTION
SHEET 13.00

3730286 CLUTCH FORK BOOT ASSY
FOR INSTALLATION SEE SECT 6
SHEET 13.00 VIEW C FOR LD
SHEET 14.00 VIEW C FOR RD

3713902 REMOVED

OIL LEVEL GAUGE
INSTRUCTION
SHEET 6.00

FILTER CAVITY INSTRUCTION
SHEET 6.01

ACCELERATOR CONTROL INSTRUCTION
SHEET 10.00 LD
SHEET 11.00 RD

EXHAUST SYSTEM INSTRUCTION
SHEET 5.00

CARBURETOR & FUEL PUMP
INSTRUCTION
SHEET 8.00

ENGINE ASSEMBLY
3836795 LD ┐3 SPEED
3836796 RD ┘
3836797 LD OVER DRIVE
3836799 LD AUTO. TRANS.

SPARK & CHOKE
LINE INSTRUCTION
SHEET 9.00

FUEL LINE INSTRUCTION
SHEETS 9.00 & 9.01

AIR CLEANER INSTRUCTION
SHEET 4.00

CARBURETOR TO TRANSMISSION
LINKAGE AUTO. TRANS INSTRUCTION
SHEET 18.00
SHEET 19.00

CLUTCH HOUSING BRACKET &
STARTER MOTOR INSTRUCTION
SHEET 12.00

ENGINE VENTILATION INSTRUCTION
SHEET 7.00

HOOD EMBLEM & ORNAMENT
INSTRUCTION
SHEET 16.00

RADIATOR INSTRUCTION
SHEET 2.00

GENERATOR, OIL FILLER, PULLEY & FAN BELT
INSTRUCTION
SHEET 3.00

WIRING INSTRUCTION
SHEET 14.00 LD
SHEET 15.00 RD

V-8 ENGINE INSTRUCTION

MODELS
ALL

		NAME	PASSENGER CAR INSTRUCTION MANUAL			SHEET
		RH.	DRAWN	CHECKED	R.P.O.	
		F		F	921	1.00
		DATE	7-25-55	PART No.	3726600	
3713902 REMOVED	T					
REDRAWN	8	5495				
REVISION RECORD		AUTH.	DR. CK.			
DATE	SYM.					
3-2-56						

13

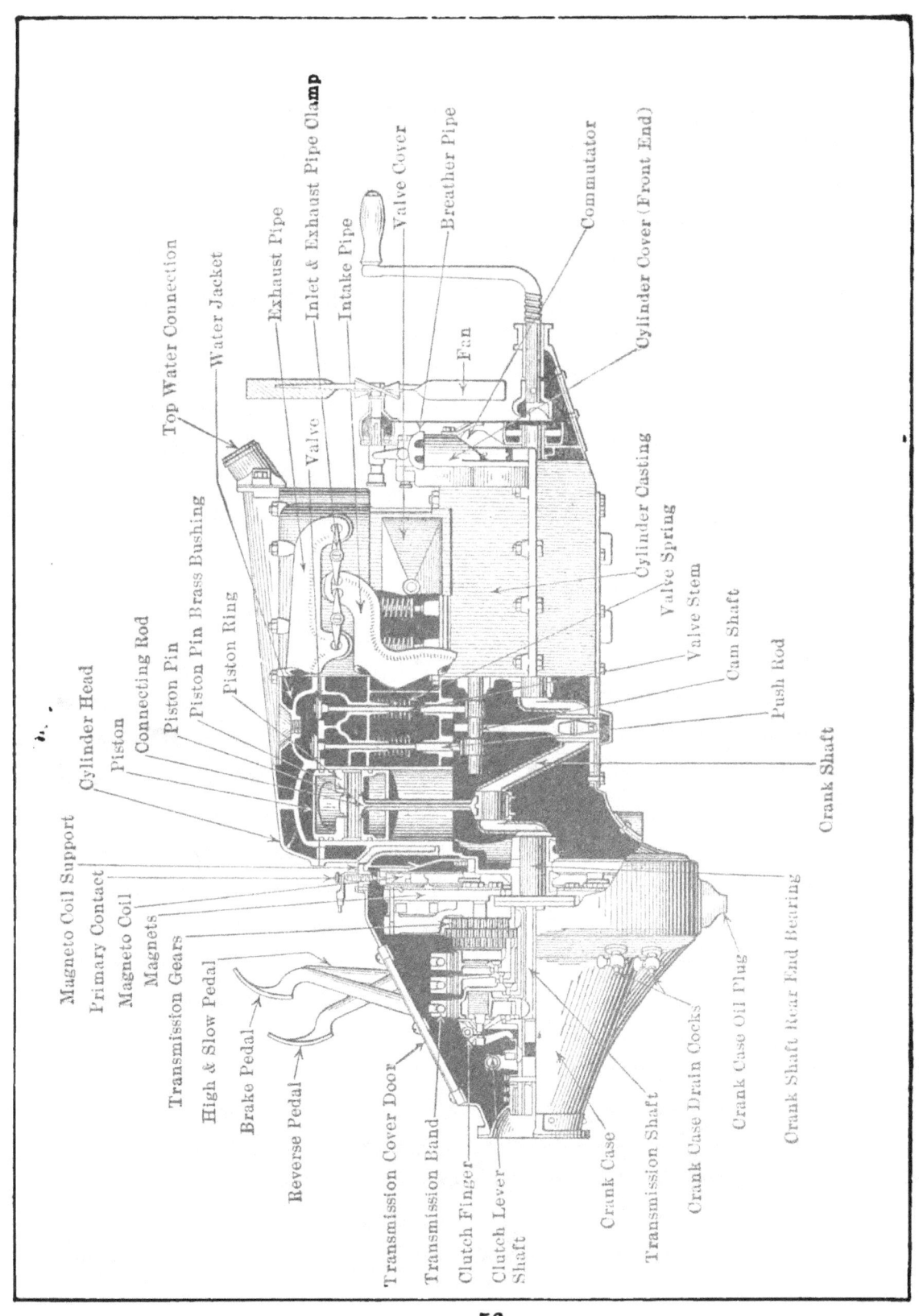

Magneto Coil Support
Cylinder Head
Piston
Connecting Rod
Piston Pin
Piston Pin Brass Bushing
Piston Ring

Top Water Connection
Water Jacket

Exhaust Pipe
Inlet & Exhaust Pipe Clamp
Intake Pipe
Valve Cover
Breather Pipe

Valve

Fan

Commutator

Cylinder Cover (Front End)

Cylinder Casting
Valve Spring
Valve Stem
Cam Shaft
Push Rod

Crank Shaft

Magneto Coil Support
Primary Contact
Magneto Coil
Magnets
Transmission Gears
High & Slow Pedal
Brake Pedal
Reverse Pedal

Transmission Cover Door
Transmission Band
Clutch Finger
Clutch Lever Shaft
Crank Case
Transmission Shaft
Crank Case Drain Cocks
Crank Case Oil Plug
Crank Shaft Rear End Bearing

Fig. 11.—Part Sectional View of the Ford Four Cylinder Unit Power Plant Showing Important Parts of the Power Generating and Transmission System.

56

15

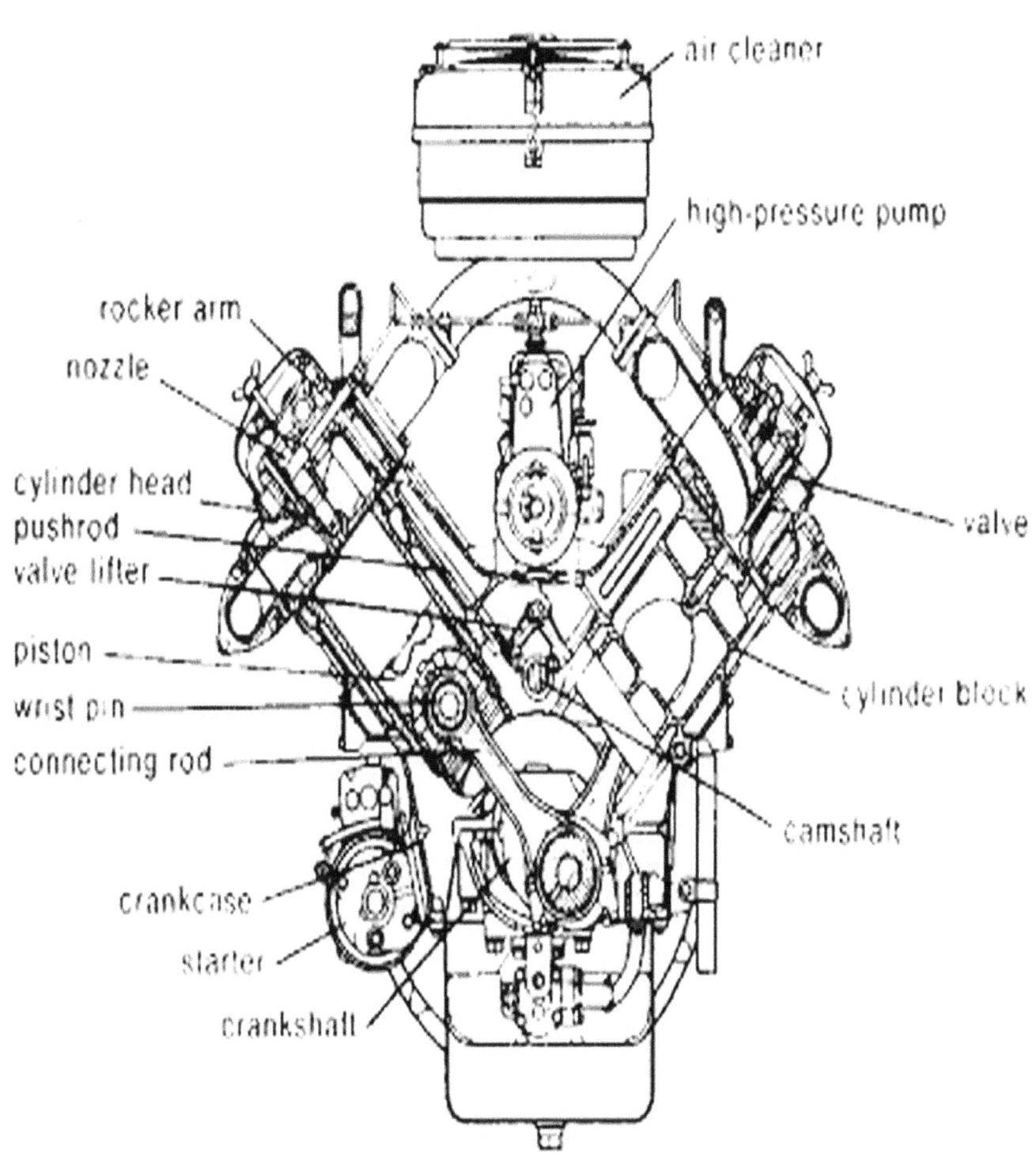

air cleaner

high-pressure pump

rocker arm

nozzle

cylinder head
pushrod
valve lifter

piston

wrist pin

connecting rod

crankcase

starter

crankshaft

valve

cylinder block

camshaft

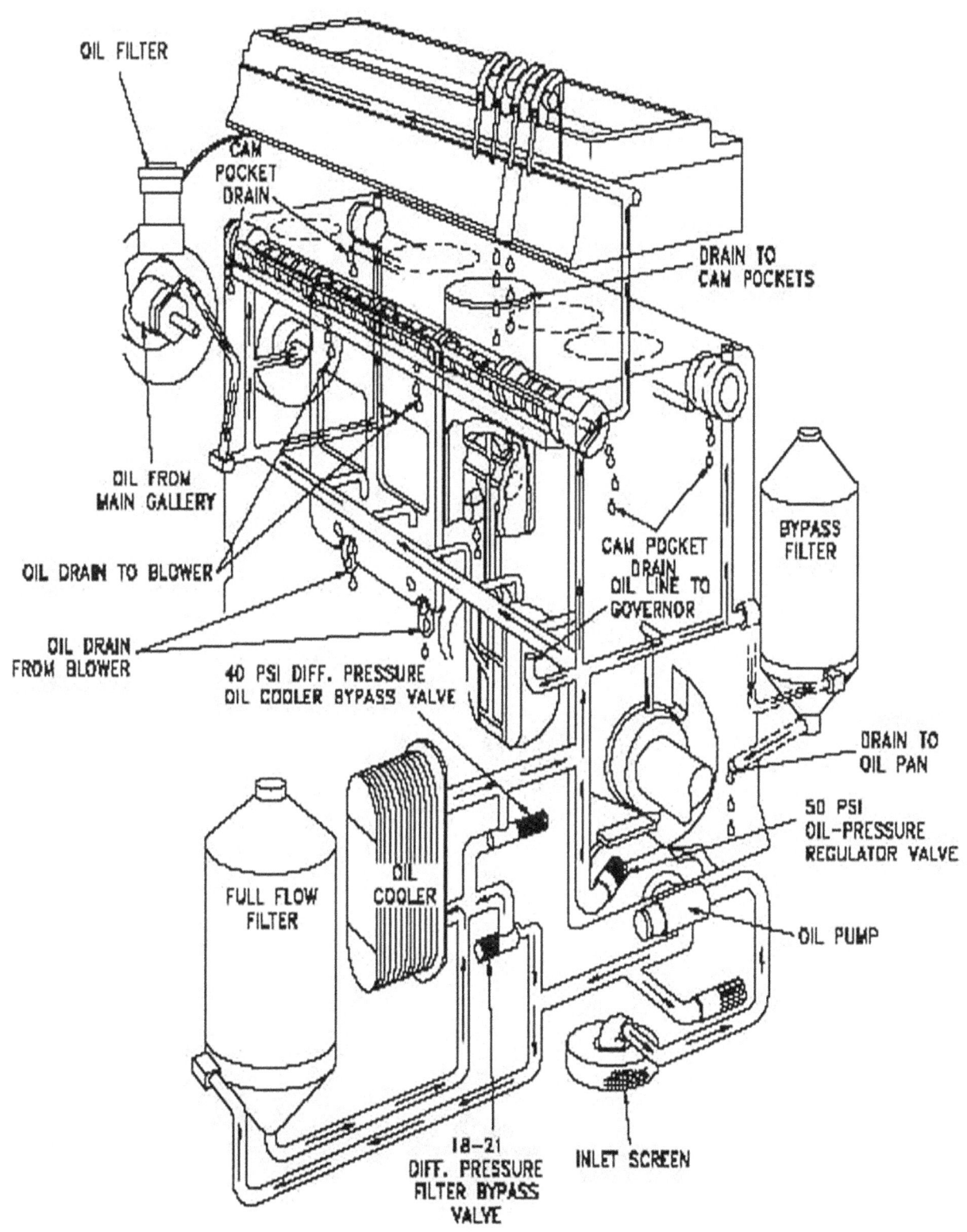

OIL FILTER

CAM POCKET DRAIN

DRAIN TO CAM POCKETS

OIL FROM MAIN GALLERY

OIL DRAIN TO BLOWER

OIL DRAIN FROM BLOWER

CAM POCKET DRAIN OIL LINE TO GOVERNOR

BYPASS FILTER

40 PSI DIFF. PRESSURE OIL COOLER BYPASS VALVE

DRAIN TO OIL PAN

50 PSI OIL-PRESSURE REGULATOR VALVE

FULL FLOW FILTER

OIL COOLER

OIL PUMP

18-21 DIFF. PRESSURE FILTER BYPASS VALVE

INLET SCREEN

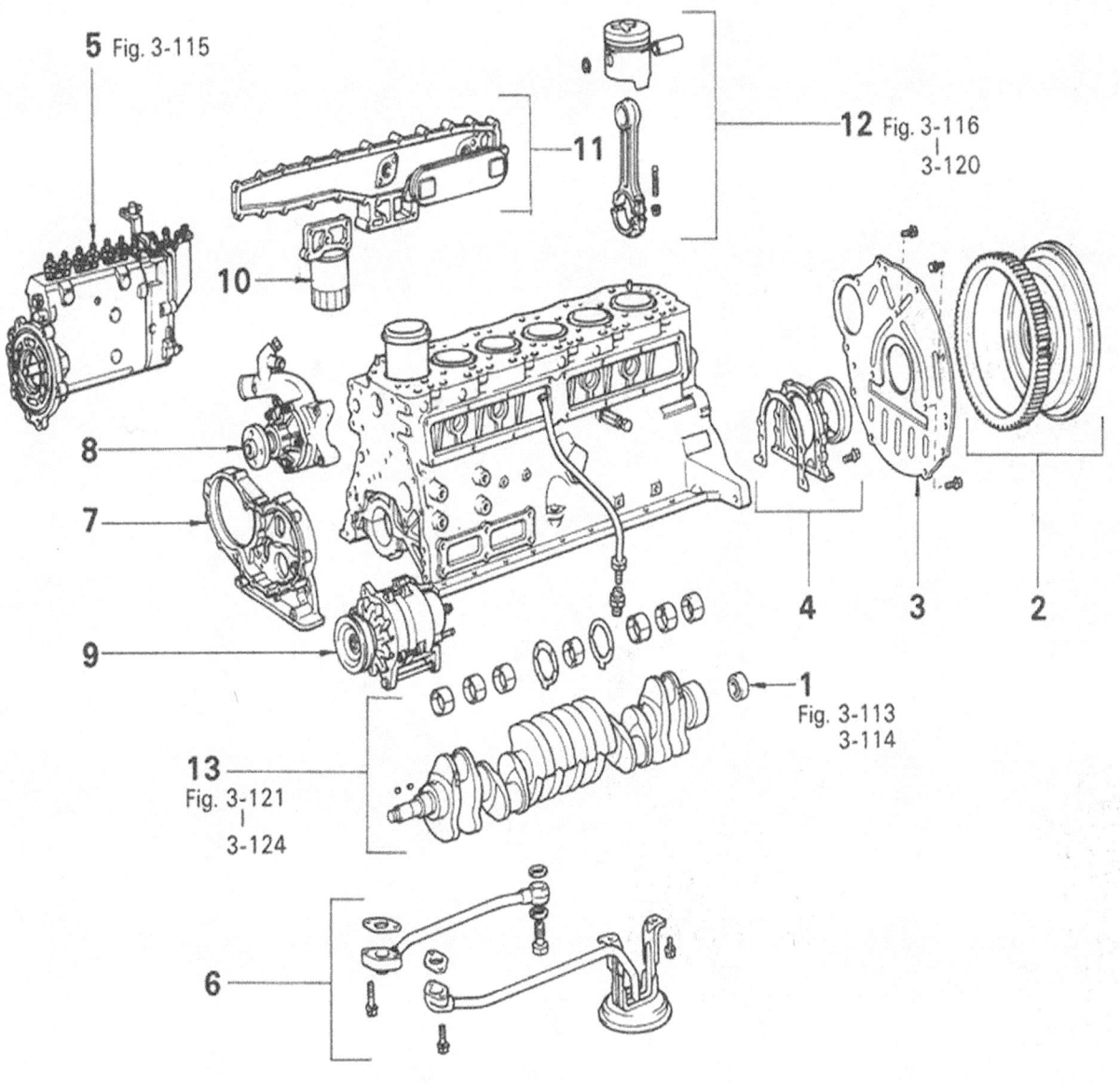

5 Fig. 3-115

11

10

12 Fig. 3-116
3-120

8

7

9

4   **3**   **2**

1
Fig. 3-113
3-114

13
Fig. 3-121
|
3-124

6

1. Input Shaft Bearing
2. Flywheel
3. Rear End Plate
4. Rear Oil Seal & Retainer
5. Injection Pump with Pump Retainer
6. Oil Strainer & Pipe
7. Timing Gear Case

8. Water Pump
9. Alternator
10. Oil Filter & Cap
11. Water Inlet Housing & Oil Cooler
12. Piston & Connecting Rod
13. Crankshaft

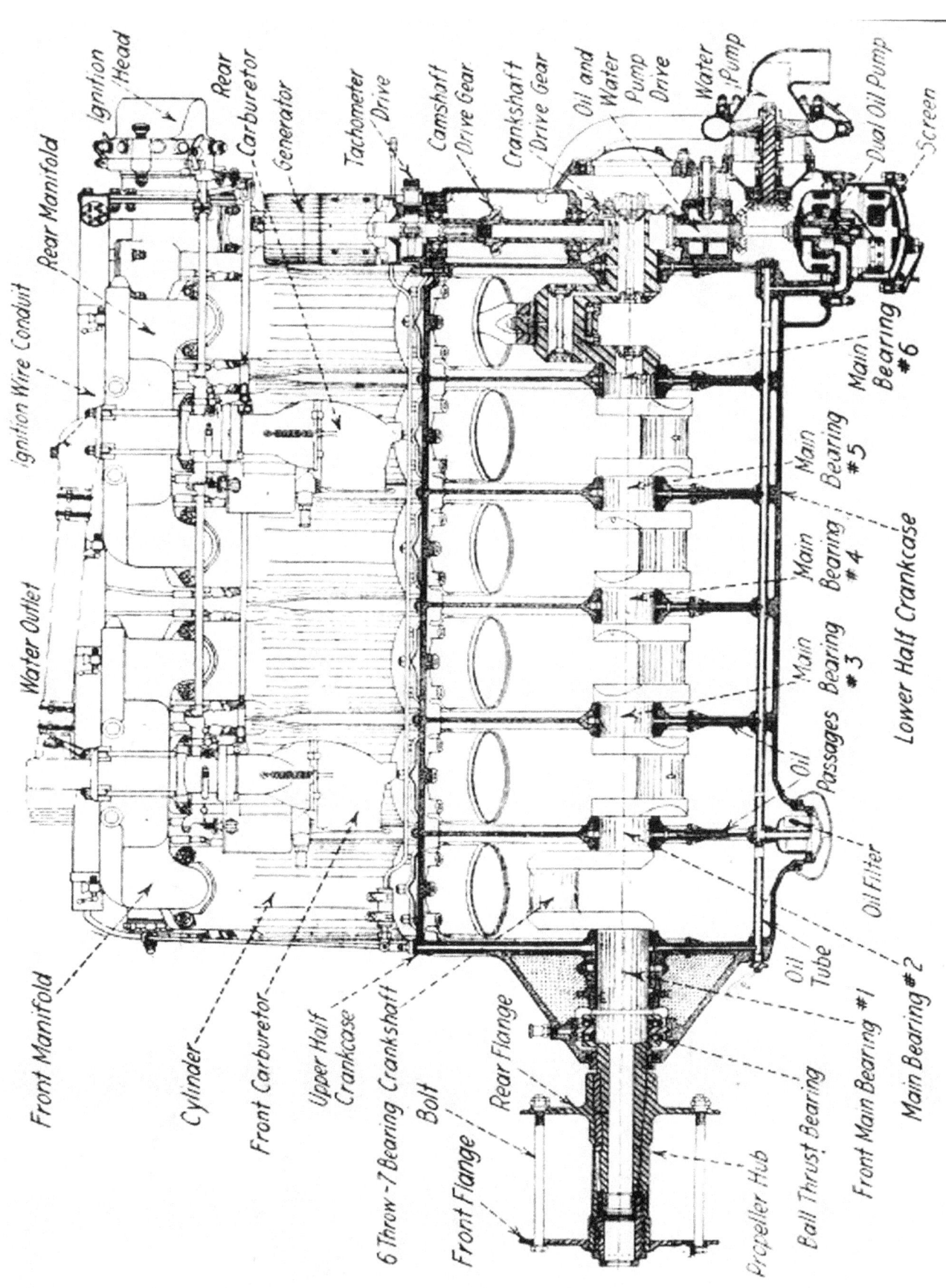

Ignition Head

Ignition

Rear Carburetor

Rear Manifold

Generator

Tachometer Drive

Camshaft Drive Gear

Crankshaft Drive Gear

Oil and Water Pump Drive

Water Pump

Water Pump

Dual Oil Pump

Screen

Ignition Wire Conduit

Main Bearing #6

Main Bearing #5

Main Bearing #4

Water Outlet

Main Bearing #3

Lower Half Crankcase

Oil Passages

Oil Filter

Front Manifold

Cylinder

Front Carburetor

Upper Half Crankcase

6 Throw – 7 Bearing Crankshaft

Bolt

Rear Flange

Front Flange

Propeller Hub

Ball Thrust Bearing

Front Main Bearing #1

Main Bearing #2

Oil Tube

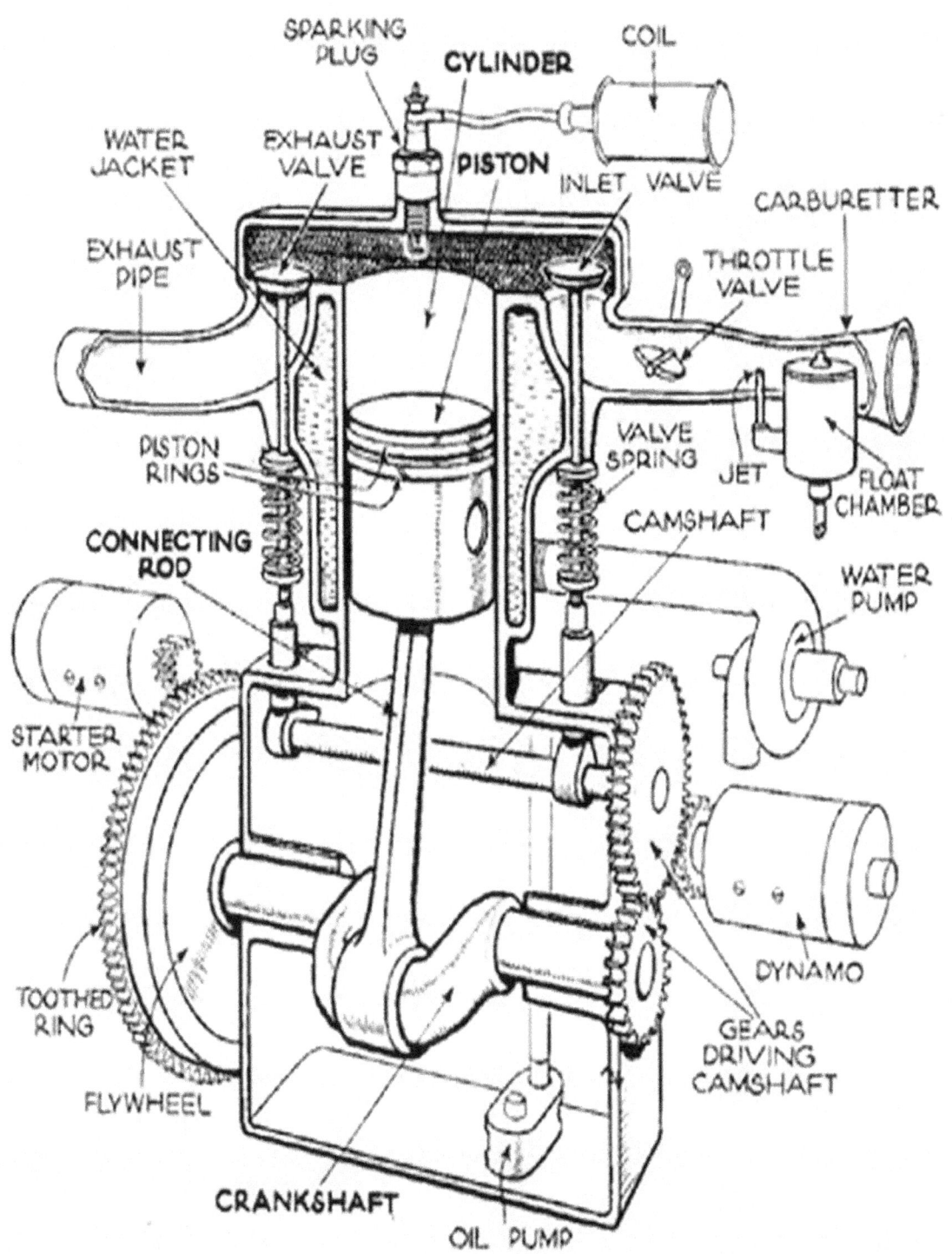

main parts of engine

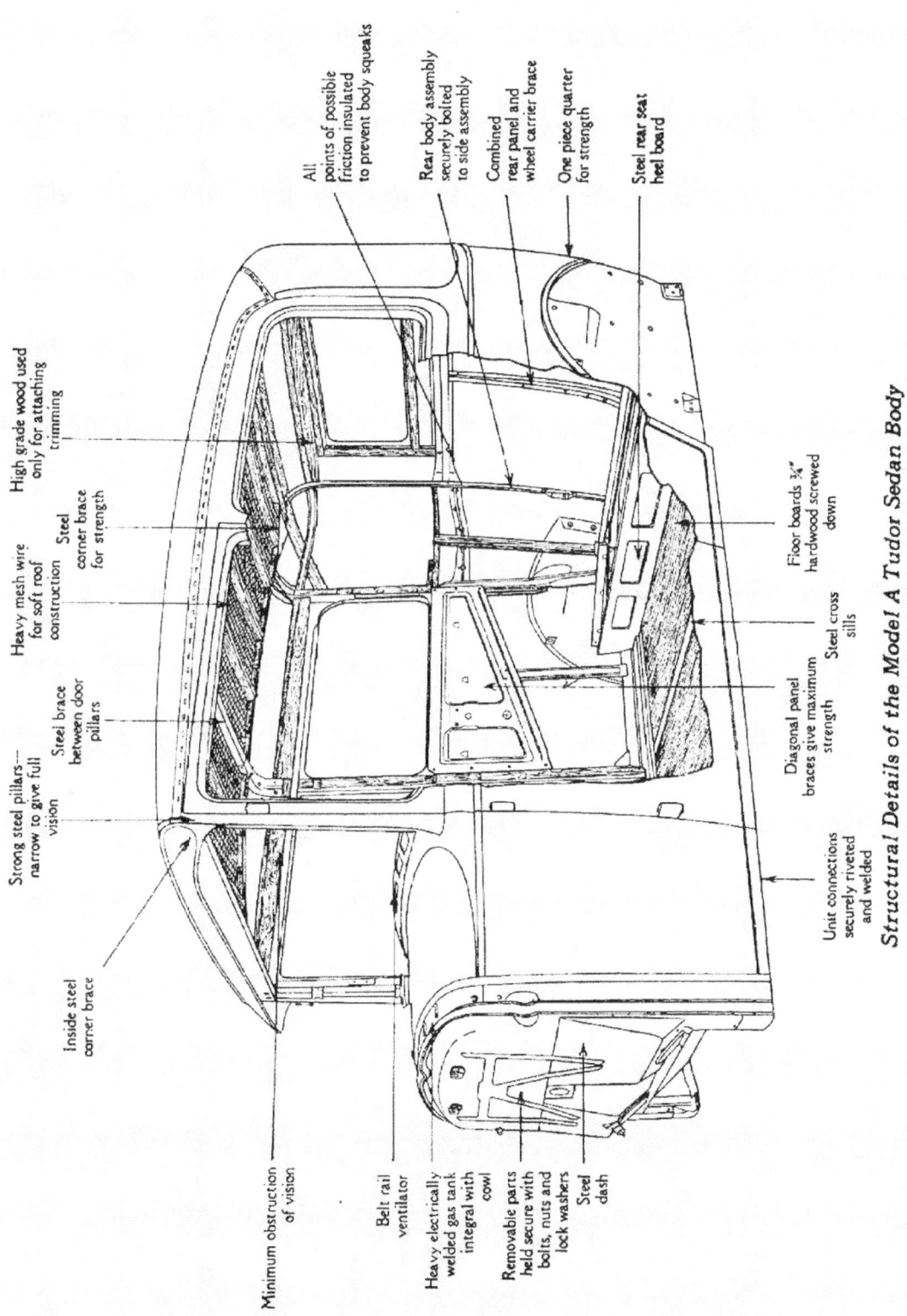

All points of possible friction insulated to prevent body squeaks

Rear body assembly securely bolted to side assembly

Combined rear panel and wheel carrier brace

One piece quarter for strength

Steel rear seat heel board

High grade wood used only for attaching trimming

Steel corner brace for strength

Heavy mesh wire for soft roof construction

Steel brace between door pillars

Strong steel pillars—narrow to give full vision

Inside steel corner brace

Minimum obstruction of vision

Belt rail ventilator

Heavy electrically welded gas tank integral with cowl

Removable parts held secure with bolts, nuts and lock washers

Steel dash

Unit connections securely riveted and welded

Diagonal panel braces give maximum strength

Steel cross sills

Floor boards ¾" hardwood screwed down

Structural Details of the Model A Tudor Sedan Body

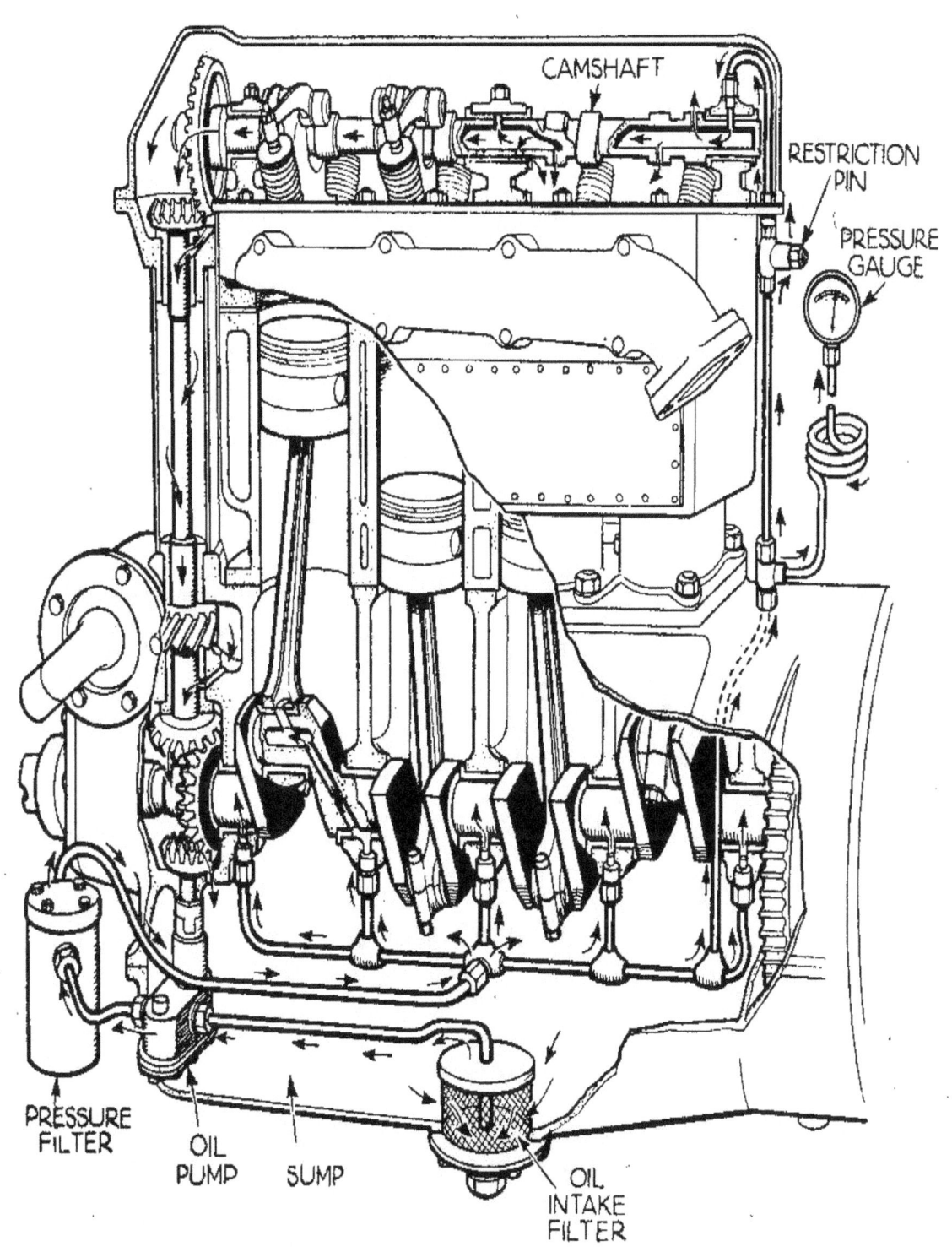

CAMSHAFT

RESTRICTION PIN

PRESSURE GAUGE

PRESSURE FILTER

OIL PUMP

SUMP

OIL INTAKE FILTER

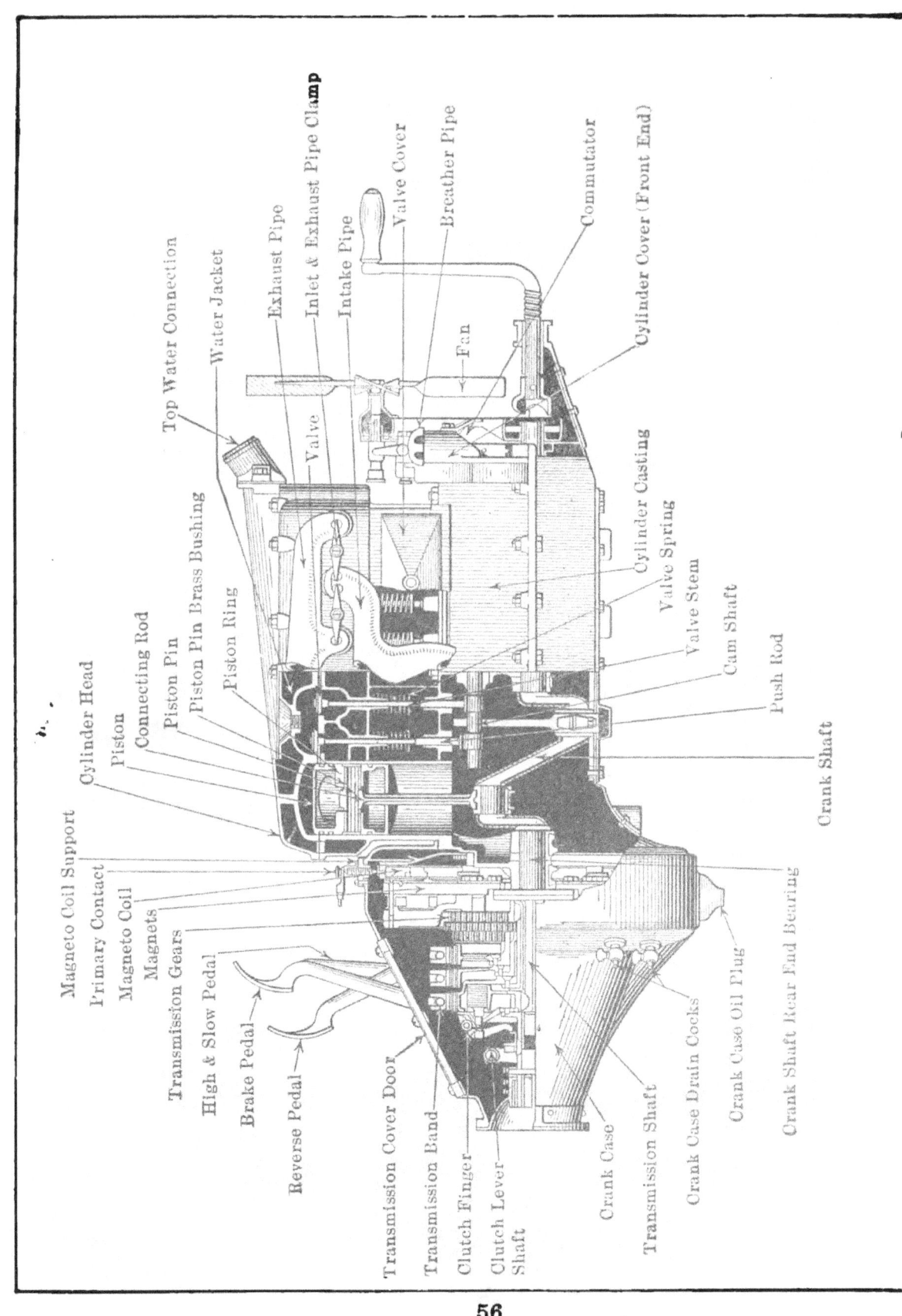

Magneto Coil Support
Primary Contact
Magneto Coil
Magnets
Transmission Gears
High & Slow Pedal
Brake Pedal
Reverse Pedal

Cylinder Head
Piston
Connecting Rod
Piston Pin
Piston Pin Brass Bushing
Piston Ring

Top Water Connection
Water Jacket

Valve

Exhaust Pipe
Inlet & Exhaust Pipe Clamp
Intake Pipe

Valve Cover
Breather Pipe

Fan

Commutator

Cylinder Cover (Front End)

Cylinder Casting
Valve Spring
Valve Stem
Cam Shaft
Push Rod

Crank Shaft

Transmission Cover Door
Transmission Band
Clutch Finger
Clutch Lever Shaft
Crank Case
Transmission Shaft
Crank Case Drain Cocks
Crank Case Oil Plug
Crank Shaft Rear End Bearing

Fig. 11.—Part Sectional View of the Ford Four Cylinder Unit Power Plant Showing Important Parts of the Power Generating and Transmission System.

56

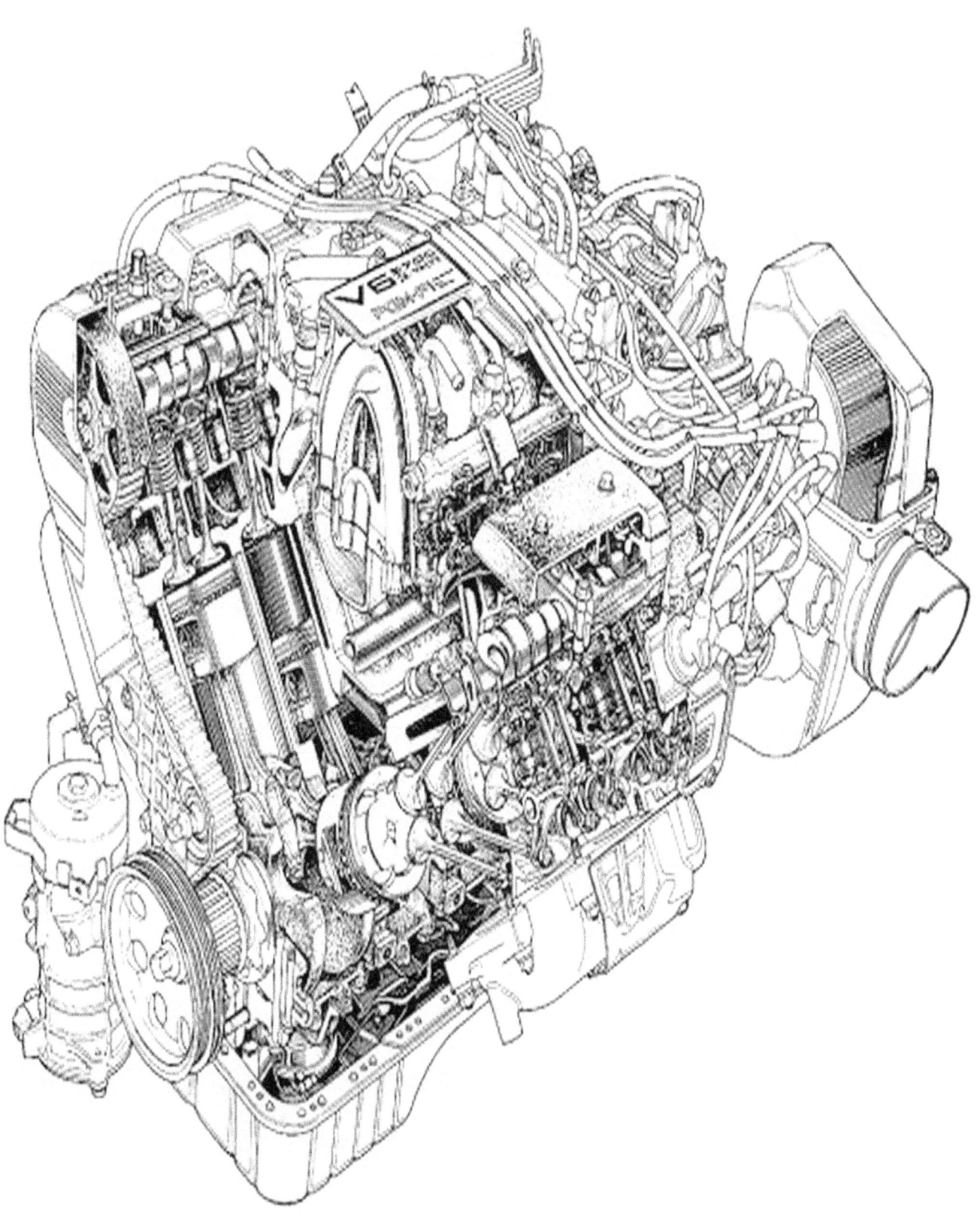

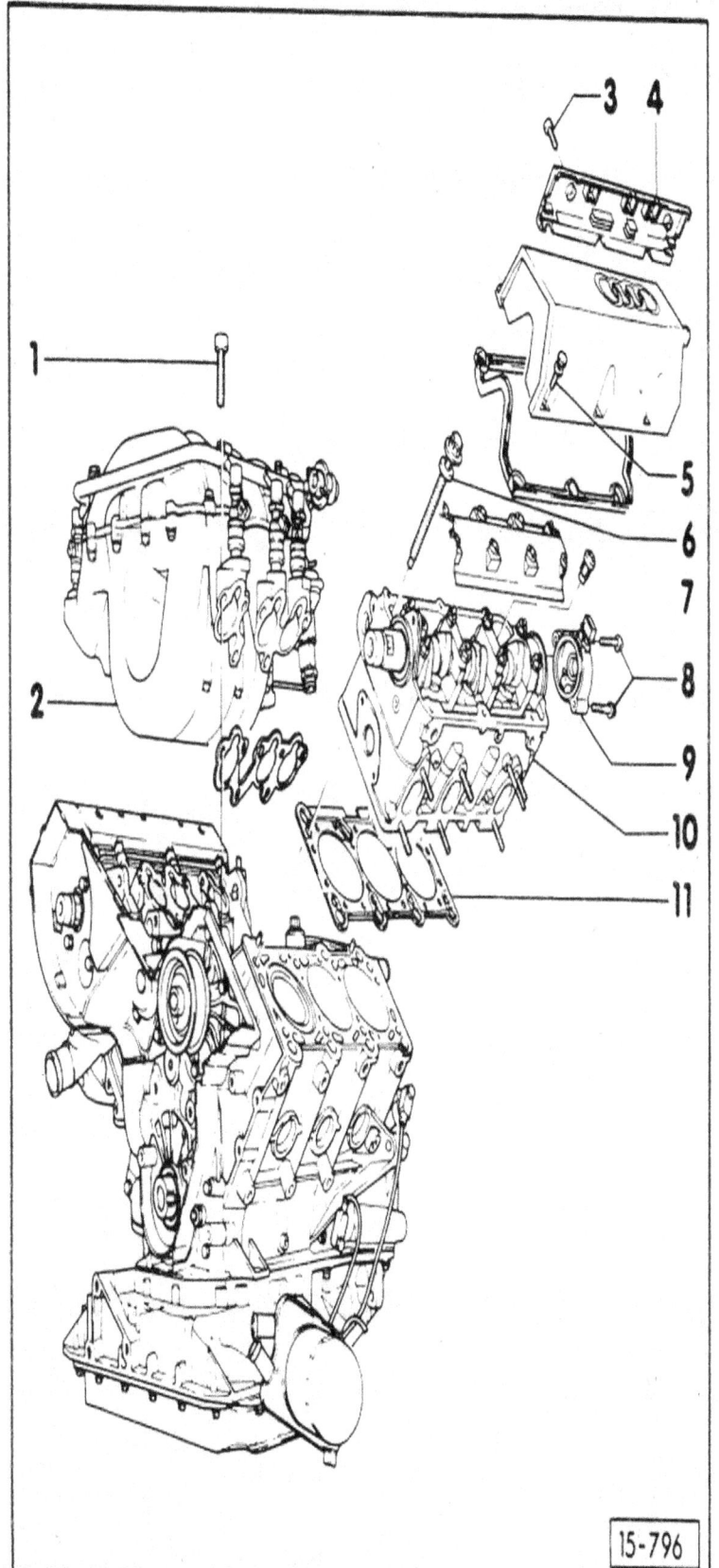

15-796

Zylinderkopf

Hinweis:

♦ *Zylinderkopfhaube einbauen*
  ⇒ *Seite 15-5*

1 - 20 Nm

2 - Saugrohr
  ♦ aus- und einbauen
    ⇒ Seite 15-12

3 - 10 Nm

4 - Abdeckung

5 - 10 Nm

6 - Zylinderkopfschrauben
  ♦ Anziehmethode
    ⇒ Seite 15-14
  ♦ Schrauben ersetzen.

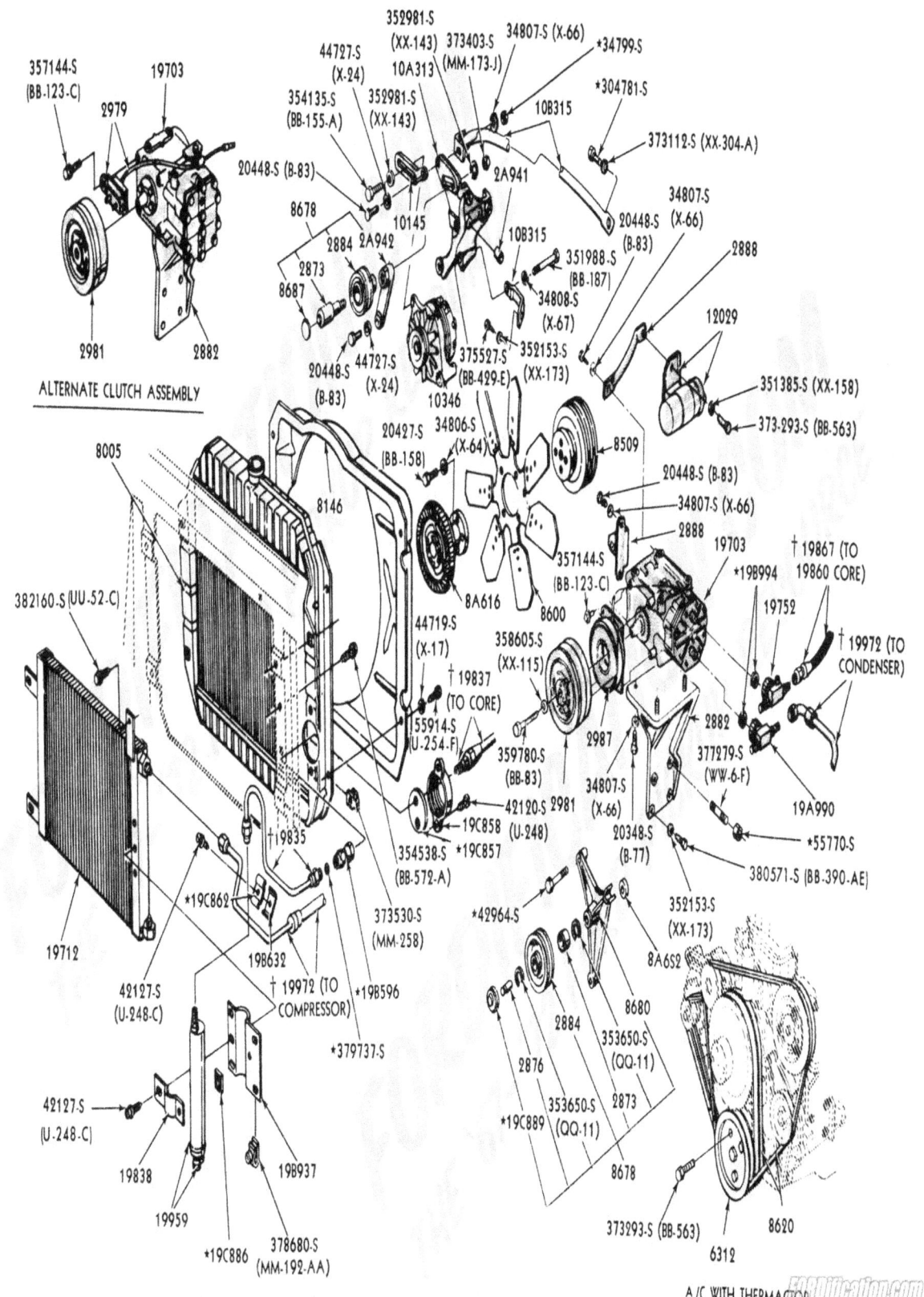

357144-S
(BB-123-C)

19703

2979

2981   2882

ALTERNATE CLUTCH ASSEMBLY

352981-S
(XX-143)   34807-S (X-66)
44727-S   373403-S   *34799-S
(X-24)   10A313 (MM-173-J)
354135-S   352981-S   *304781-S
(BB-155-A)   (XX-143)   10B315
20448-S (B-83)   373112-S (XX-304-A)
8678   2A941
10145   34807-S
2884 2A942   10B315   (X-66)
2873   351988-S   20448-S
8687   (BB-187)   (B-83)
20448-S   44727-S   34808-S
(B-83)   (X-24)   (X-67)
375527-S   352153-S
(BB-429-E)   (XX-173)
10346
2888
8146   20427-S   34806-S   12029
(BB-158)   (X-64)   351385-S (XX-158)
8005   373-293-S (BB-563)
8509
20448-S (B-83)
382160-S (UU-52-C)   34807-S (X-66)
2888
357144-S
(BB-123-C)   19703 *19B994   † 19867 (TO
8A616   19860 CORE)
44719-S   8600   19752
(X-17)   † 19972 (TO
358605-S   CONDENSER)
(XX-115)
† 19837
(TO CORE)
55914-S   2882
(U-254-F)   377279-S
359780-S   2987 (WW-6-F)
(BB-83)
42120-S 2981   19A990
19C858   34807-S
354538-S *19C857   (X-66)   *55770-S
(BB-572-A)   20348-S
373530-S   (B-77) 380571-S (BB-390-AE)
19712   (MM-258)
19B632   352153-S
*19B596   (XX-173)
42127-S † 19972 (TO   8A6S2
(U-248-C) COMPRESSOR)
*379737-S
42127-S   8680
(U-248-C)   *42964-S   353650-S
(QQ-11)
2884
19838   2876 353650-S 2873
19B937   *19C889 (QQ-11)
19959   8678
378680-S   373293-S (BB-563)   8620
*19C886 (MM-192-AA)   6312

A/C WITH THERMACTOR
P-6447

† REFER TO TEXT SECTION 197 FOR SERVICE REFRIGERENT LINE ASSY. and/or REPLACEMENT DETAILS

AIR CONDITIONING PARTS – ENGINE COMPARTMENT
1968/72 F100/350 – 6 CYL. 240, 300 – w/INTEGRAL A/C

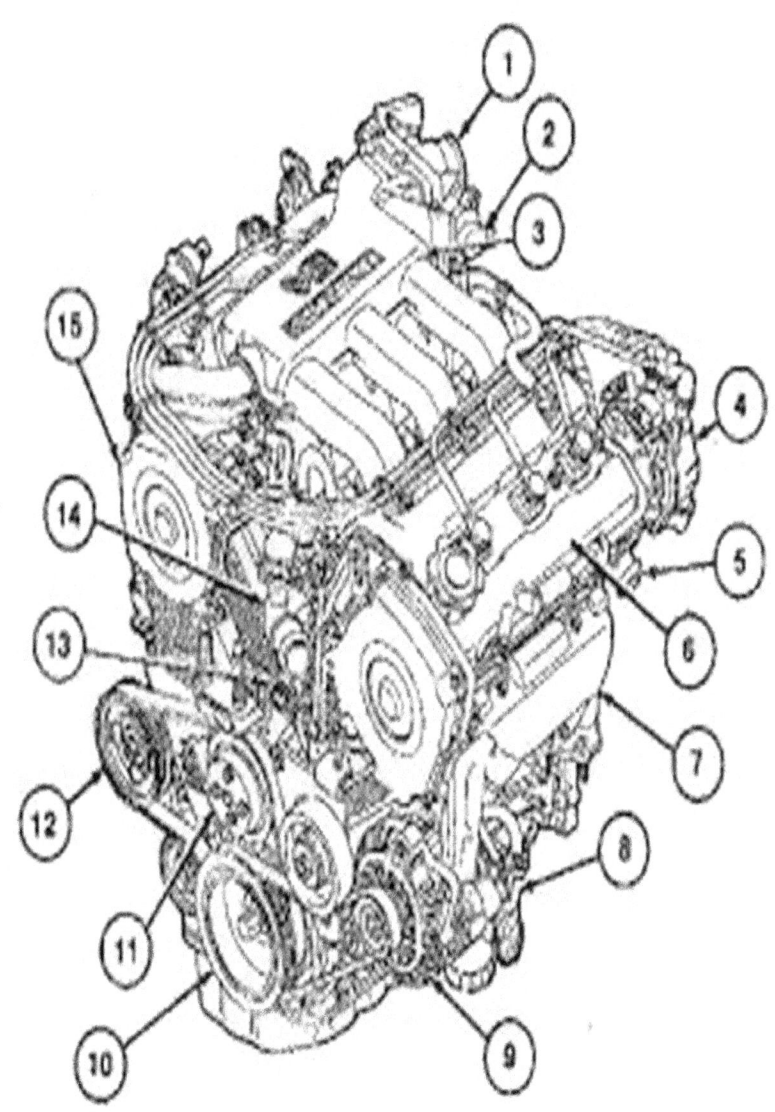

Thank You For Coloring N.H.M.P.

www.ingramcontent.com/pod-product-compliance
Lightning Source LLC
Chambersburg PA
CBHW081134180526
45170CB00008B/3099